This book was devised and produced by Multimedia Publications (UK) Ltd.

Editor: Jeff Groman
Production: Arnon Orbach
Design: Brian Harris
Photography: Nicky Wright

First published in the United States of America 1985 by Gallery Books, an imprint of W.H. Smith Publishers Inc., 112 Madison Avenue, New York, NY 10016

ISBN 0 8317 1163 9

Origination by D S Colour International Ltd., London
Printed in Italy by New Interlitho SpA, Milan

Dedication for Sharon and Julian with love

A fully optioned '81 Z-28
a-rarin' to go.

CAMARO

Watch out, worms! That deep air dam will decapitate you! However, the IROC-Z's four wheel disc brakes are powerful, straight and sure in an emergency.

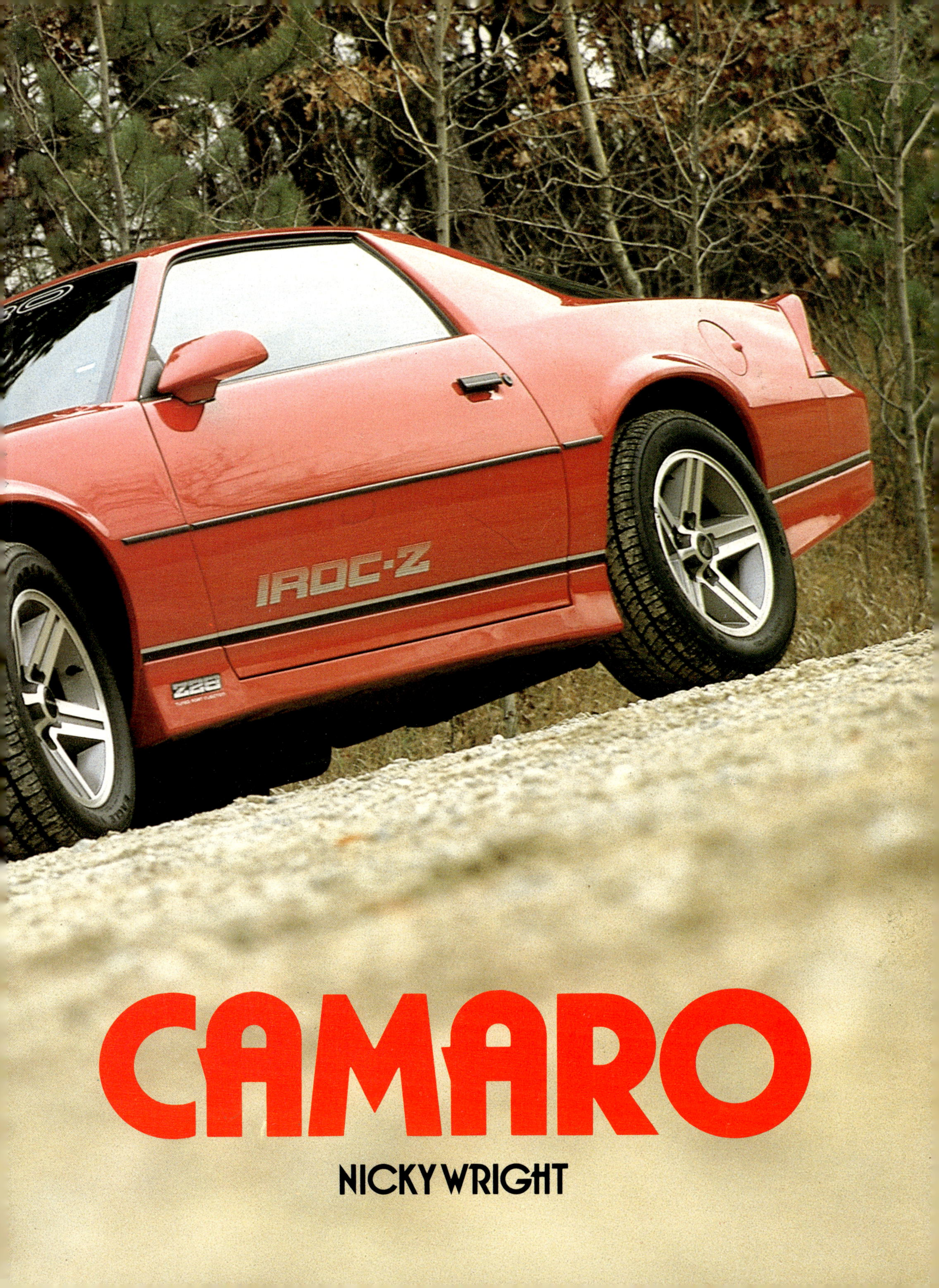

CAMARO

NICKY WRIGHT

The bold stripe tells us this is a 1967 SS-350 Camaro and unique to this model. A notch below the Z-28, the SS-350 could turn a creditable 0-60 in 7.8 seconds. (General Motors)

CONTENTS

INTRODUCTION

April 1964 saw the birth of a new trend in motorcar design. It was Ford's Mustang, the first of the Ponycars, (so described because of its relatively small size but sporting nature). The Mustang's success is history now, a concept that survives to this day in Chevrolet's Camaro.

The Camaro arrived on the scene 2½ years behind the Mustang along with a herd of ponies all eager for a piece of Mustang's success. From the start it was obvious the Camaro was a special kind of car. Its formula was the same as Mustang's; a low base price and options galore to allow the buyer to 'build' his own personal standard.

It had been an uphill battle against Ford's Mustang, then deeply entrenched in the public mind as the sporty car to have. But perseverance has its rewards, and in 1974 Camaro found itself almost alone, with only its 'kissin cousin,' Pontiac's Firebird, for company.

Throughout the shaky seventies, with Detroit grinding on through the oil crisis, then into the recession-blighted eighties, the Camaro soldiered on, winning new friends, gaining in popularity. In 1982 the latest Camaro launched a new generation of high-tech super cars culminating in the world beating IROC Camaro.

Collectability assured and sales still healthy, the Camaro has emerged from the shadow of the Mustang as a car of class. After reading this book we think you will appreciate the Camaro as one of the world's truly great automobiles.

Below: This beautiful 1980 Z-28 is loaded with just about every option in the book and then some. At least the 350 V8 was standard as were the disc brakes. Left: A 1970 base Camaro in full flight at a time when speed and poor gas economy weren't dirty words. (General Motors).

TO CREATE A LEGEND

Today the Camaro has come of age. 1985 saw the arrival of the Camaro IROC, patterned on the racing Camaros which were created especially by Chevrolet for the now renowned International Race of Champions. The result of this development is that the Camaro IROC of today equals or betters exotic European sporting machinery costing two or three times as much, and is further proof (if any were needed) that racing produces better – and safer cars.

The Camaro IROC

The IROC however, is not just a straight line accelerator of old; it handles, brakes and corners with the best of them and in its price range is arguably the best. A far cry from the brute force and ignorance days of the mid to late sixties, when the cry was 'there's no substitute for cubic inches'; when growling exhausts and screaming tires tore up the 1/4 mile at track and stoplight. It was into this environment that the Camaro first saw the light of day.

General Motors (GM) had relied too much on their nifty and pretty Chevrolet Corvair Monza Spyder, complete with bucket seats and four on the floor, to take on Ford's newly introduced Mustang Ponycar. Before the Mustang and Ralph Nader's book, *Unsafe At Any Speed*, the Corvair Monza enjoyed two years of profitable, six figure sales. The Mustang crushed it like a runaway bulldozer and Nader did the rest. Chevrolet was out of the running and in August 1964 GM's No. 1 division decided to develop a 'kissin' cousin' answer to Ford's huge sales success; that is, a car built with off the shelf components and bristling with an option list to enable the

buyer to build it as he saw fit. As development got under way, GM would have to grit its corporate teeth for two years as the unstoppable Mustang paved Ford's way with gold.

The F – Car Project

According to GM sources, the idea for an inexpensive, four passenger sporty car had been considered, even worked on, six years before the Mustang made its debut. At the time corporate bosses paid little attention to what engineers and designers told them, so the idea never took root. It was only when they saw the Mustang steam-rolling everything in its path that GM hastily decided to do something about it. Four months after the Mustang launch, Chevrolet set to work on its 'Mustang buster'.

Like Mustang, the 'F-car', as it was designated early on, would rely on components already available. This would save time and, more importantly as far as the accountants were concerned, money. Body design was handled by Chevy's Studio Two under the aegis of Henry E. Hoga, though other GM designers also played a part. The great William L. Mitchell, then GM design vice-president, let it be known that he didn't like the car, considering it a case of too many cooks spoiling the broth. Styling was limited to a certain extent as it was meant to be able to take already existing components.

As it actually transpired, the F-car, by now christened Camaro, shared a lot less than originally intended. Unique to the Camaro was its front sub-frame combined with a unitary body from the cowl rearward. This format incidentally, was adapted for the 1968 Chevy Nova.

Top left and top right. Chevrolet's SS 350 Camaro dream machine replete with feminine encouragement. This was the way Madison Avenue saw the sporty car world in 1967. Above: 1967 Camaro being put through its paces. (General Motors).

Styles and options

So the Camaro was virtually a new car, relying less on the 'parts bin' than the early Mustangs. But a careful observer would find bits and pieces from the Corvair to the full size Chevrolet: the mono-plate single leaf spring was taken from the Chevy II – an arrangement that was not entirely successful due to axle tramp when combined with bigger engines. Camaro's base engine was the 230 cu in 140 bhp six as used in Chevelle; options included the larger 250 cu in six from the full size Chevrolet and a host of V8s running from a mild 302 to a whopping 396 cu in, pumping out 325 bhp.

The model line-up consisted of only a coupé and convertible; management shied away from other styles due to cost. These were offered in three series; base Camaro, Rally Sport with hideaway headlights and the macho SS 350, which came with the 295 bhp 350 cu in V8 as standard. Like the Mustang, the buyer could tailor his choice three ways: Economy, Luxury or Performance. The options list had everything from a $7.00 deluxe steering wheel to $356 worth of air conditioning.

Overall length was an inch longer than Mustang at 184.6 inches while the width was almost 2 inches more at 72.5;

most other dimensions were fairly similar. As the Camaro was destined to be a Mustang beater, it was priced the same way. Standard Camaros cost approximately $2550 up to the Performance model at around $4000. The costliest approximation would have been the Luxury version. Loaded it could set the customer back at least $4600. The average Camaro left the showroom with a $3-3500 price tag, again much the same as a similarly equipped Mustang.

The car was scheduled for a Fall 1966 launch as a 1967 model. Chevrolet had lost its traditional first place position to Ford in 1966, so the Camaro had a lot more riding on it.

Opposite and above: A 1968 Camaro SS-396 reposing in the tranquility of an early fall evening: note the 'Now you see them, now you don't headlights. The 396 engine cost extra but adds dollars on today's collector market. Top: A 1967 Camaro under full power at GM's proving grounds.

These pages: A well restored 1968 SS with a mighty 427 V8 under the hood. In its day an impressively quick automobile although axle tramp could be a problem with earlier models. Insets: Front end treatment was neat and simple, the headlamp arrangement well executed.

SS

RADIAL
STARFIRE

These pages: Although a 1969 model, this Camaro's appearance reveals minor modifications, a common practice with the go faster fraternity. This one has been tastefully done, with the removal of side lights and emblems. The chrome scoops on the hood and rear louvers were only for show.

Main picture: A 1969 modified Camaro showing its exceptionally clean lines. Inset: A British-owned SS 350 lurks in an English country lane.

SS
THX
243M

INTO THE FRAY

First year sales for the Camaro didn't exactly set the world afire, but they were healthy nonetheless. 220,717 Camaros helped Chevy overhaul Ford for the No. 1 spot in 1967, even though Mustang held its commanding lead in the Ponycar stakes with more than double the Camaro's figures at 472,121 units produced. True, Mustang sales were down over 1966 when only Plymouth's Barracuda offered a challenge. Both Camaro and Mustang faced increased competition in 1967 with the arrival of the Cougar and Firebird, Pontiac's Camaro lookalike, grille excepted. It would be interesting to speculate how well Chevrolet's contender would have done in a direct head to head confrontation with Mustang without having to score points off all the others. The consensus is that Mustang would have stayed way out in front.

Taking on Mustang

By 1969, Camaro's presence had begun to hurt Ford; the original pony's sales had fallen dramatically. Mustang still held onto its lead but the gap had narrowed to 56,000 units. Chosen as the official pace car for the Indianapolis 500 in 1967 and 1969, handsome looks and the muscular Z-28 model notwithstanding, Camaro had a lot going for it. Its wind-tunnel tested shape was rounder, more sensual than the angular Mustang styling and in SS-350 guise with its black out grille and hidden headlights, it looked mean for the macho male yet softly romantic, the stuff that girl's dreams are made of.

The Camaro met less than ecstatic reviews from the auto magazines. Everybody liked its looks but didn't think much of its handling. 'Chevrolet has had at least two years to come up with a car demonstrably superior to the Mustang,' said *Car and Driver*. 'That Chevy hasn't, may speak volumes about Ford. Maybe GM can't build a better mousetrap, only a good one'.

On the other hand, Camaro was surprisingly well received by Britain's auto press. *Motor* magazine ended its test with the aside that it was 'one of the best American cars we have driven', while William Boddy, *Motor Sport* magazine's most durable fixture, although observing it might be too big for London parking, was generous in his opinion that it was the best sports-type car he had experienced from across the Atlantic. Considering the usual anti-American bias that is the forté of most British car pundits, this was high praise indeed!

The cost of safety

As mentioned earlier, the Camaro's major design fault was its rear axle and single leaf spring suspension. This was just passable with the basic six but right out of its league with a V8. Its short-comings showed up time and again in road

These pages: This particular 1968 Z-28 was bought new by its owner and is about the most perfect example of an early Zee to be seen. Originally conceived for competing in the Trans-Am race series, Chevrolet meant only to make enough for homologation purposes. But demand was so great, 7199 units were made in 1968, 19,014 in 1969.

tests, drivers complaining of extreme axle judder accompanied with a disconcerting habit to hop sideways when a heavy right foot was applied. Chevy engineers were quick to try and remedy the problem with traction bars and staggered shocks, which reduced but didn't eliminate the problem. So the ignominious mono-plate springs were eventually thrown away, to be replaced for 1970 by semi-elliptic multileaf suspension. This resulted in a vastly improved car. One might be given to question the wisdom of the original single leaf set up. The answer lies not at the engineers' door, but in the hands of the cost cutters to whom safety and handling take second place to profitability.

The ultimate Camaro. Part I

The Z-28 was originally conceived as a $437.10 option for the stock 1967 Camaro. The Z-28 designation referred to a 302.4 cid., 290 bhp V8 engine, heavy duty front and rear springs and shocks, 15 x 6 in wheels and racing stripes. The engineers took a 283 crankshaft, put it into a 327 block, and added mechanical lifters in place of the standard hydraulic ones for better and higher revs-per-minute capability. In essence they had created a thinly disguised racing engine. Mindful of the poor reception the Camaro's rear suspension had got, Chevy engineers stuck one traction arm on the right side of the car ... why only one only the motoring gods would know. With the result that axle tramp was more or less eliminated on right hand corners but not on the left ...

The reason for the Z-28 option was two-fold. One for the public at large and two, to homologate the car for the Sports Car Club of America's sedan races. Hence the 302 engine, the SCCA drew a line at 305.1 cu in. Another advantage, already considered by Chevrolet, was that the Z-28 was the ideal car to pit against Shelby's mighty Mustang GT 350 and GT 500. Performance figures between a 428 cu in powered Shelby GT 500 and the 302 Z-28 say something for the Chevrolet engine, for there was only a .2 difference in 0.60 acceleration times; the Camaro took 6.7 seconds versus 6.5 for the Shelby. From 0-80 and up however, it was Z-28 all the way. Even the ¼ mile run, found the Z-28 almost a ¼ second faster. To be fair, the Shelby carried almost 300 lb extra weight, but a test on the more comparable Shelby GT 350 weighing just 26 lb more, showed the Z-28s superiority all down the line. Here indeed was a car from which great things could be expected.

Racing Zee

602 Z-28 packages were sold in 1967, meeting the minimum quantities required under FIA (Federation International d'Automobile) rules. This doesn't seem very many, but once word got out that Chevy had made a stormer out of the

Camaro the orderbooks were filled to overflowing, with 7000 Z-28 sold in 1968, some 6600 more that Chevrolet had bargained for. The Z-28, in a nutshell, showed what could be done with a fairly straightforward car costing around $2500. Add the Z-28 package for $437, throw in power assisted disc front brakes and one of three 4 speed manual transmissions – both options were almost mandatory with the Z-28 – for an additional $42.15 and the buyer had a lot of car for a mere $3300.

Since the Z-28 was intended for racing, it wasn't long before it did just that. Most, if not all of the 1967 Z-28s went to race drivers such as Mark Donohue and Roger Penske. Others went to race orientated dealerships such as Nickey Chevrolet in Chicago and Don Yenko in Pennsylvania. The cars were prepared, tested and raced in Trans-Am competition but various shortcomings, notably in handling and braking, left the Camaros very much the underdogs at the end of the 1967 season. Not too happy with its less than auspicious showing, Chevrolet's Research and Development division took a decisive hand in turning the Z-28 into a winner.

Modify to win

Numerous modifications were carried out on the Z-28, modifications that found their way into standard production Camaros. Using one of the problematical Penske cars as a guinea pig, and driven by Donahue at GM's Milford proving grounds, Chevrolet's engineers, using a range of sophisticated equipment, spotlighted the weak points and eradicated them one by one. The result was a stronger body, multi-leaf rear suspension and better brakes. Back on the track the improvements paid off and the Z-28 saved a little face by winning the last two 1967 races.

1968 and 1969 were good years for the Z-28. It might have lost the first two races of the 1968 season but it convincingly won the Trans-Am competition, winning the nine remaining races. It repeated its success the following year, taking 8 out of the 12 Trans-Am events. Camaro was proving to the motoring world that it was a force to be reckoned with.

Above left: A mint example of a '68 Zee. Above right: A factory-produced replica of Camaro's 1969 Indianapolis Pace Car. Based on the SS 350 model, the Indy car had trim features exclusive to itself.

SS

Opposite page: The 1969 Camaro SS Indy Pace Car. Attractive interior was unique to Pace car replica. This page: A fine 1969 Camaro Rally Sport (RS) 350. Fake louvers and mag type wheel covers were part of the RS package optional on all models except Z-28. Headlight covers had three glass sections to allow some illumination if covers failed to retract.

Second generation Camaro appeared in 1970 with all new body which stayed in production for eleven seasons. Model line up, engine options much as the first generation. Car shown is a 1972 Sport Coupé with optional 350 V8, vinyl roof and trim package.

1974

Although the body remained the same, forward facing bumper-bisected grille was replaced by sloping front end. Thicker bumper complied with Federal 5 mph impact regulations. (General Motors)

Above: 151,008 Camaros were produced in 1974 at a time when most other manufacturers were pulling out of the muscle/pony car stakes. Right: Even with emissions, less power, 1974 Z-28 still meant business. (General Motors)

Z28

DIGGING IN

The first generation Camaro had been designed in a bit of a rush, no doubt to capitalize on the Ponycar market created by Mustang. Not so the second generation model which appeared in 1970. Work began on its development immediately after the first car had been finalized for production.

Past lessons

Not that the original Camaros had been all that bad. To be sure they had serious faults, as mentioned earlier, but considering the time allotted for development they had a lot going for them. With more time to create a totally new car, Chevrolet engineers and designers gave it their all.

That the new Camaro was so handsome and technically superior was in part due to William L. Mitchell. Mitchell, as head of GM Design, has been responsible for some of the best looking cars ever to come out of Detroit: the 1963 Riviera and Corvette Stingray are cases in point. Although he had little to do with the first Camaro, a car he contemptously dismissed 'as not worth a damn', the second generation Camaro has his influence written all over it.

Unique for 1970

Mitchell was able to convince GM executives that the Camaro was selling well enough to justify it having its own body dies. Therefore the new model would be unique, sharing none of its body sections with other cars in the Chevrolet line up. It goes without saying that Mitchell got his

way, freeing the designers from the restrictions imposed on them in the design of the 1967-9 Camaros and allowing them to develop the 1970 model from the ground up.

The result was an extremely attractive automobile with handling to match. The front end was expensive looking, European in concept, with an egg-crate grille similar to one of the classic Ferrari designs of the fifties. Headlights were in the fenders and no longer part of the grille, as before.

Convertibles had fallen from favor, so the new Camaro was offered in the coupé style only. It could either be the basic Sport Coupé, the Rally Sport (RS), Super Sport (SS) or finally the Z-28. The options list was as long as ever and included special interior fittings such as woodgrain appliques on the dash panel and steering wheel.

New mechanics

Mechanically the cars were a great improvement over the old. The rear suspension was revised to include staggered shocks, new spring bushings – the latter to eliminate lateral shake and body roll. Front disc brakes became standard and the steering was redesigned, the SS and Z-28 models having 'high effort' steering for more positive response.

On the engine front, the 302 was dropped in the Z-28 to be replaced by a special 350 cid., V8 producing 360 bhp. Base engine was the 155 bhp 250 cid., six followed by three 350 units in various stages of tune. Two 396 cid., V8s completed the picture. Due to increasing government regulation and emissions controls, the shattering performance exhibited by the earlier Camaros was soon to be a thing of the past. Emphasis instead was placed on refinement coupled with a quieter and more luxurious ride. For all that, the 1970 Camaro was no slouch, with 0-60 times

Top left: 1970 Camaro Sport Coupé shows inspired design that remained more than a decade in production. Above top: 1970 SS shows its heels, a view not uncommon to other drivers. Above: The '70-'73 front has distinctly European parentage. (General Motors)

of 8.8 seconds for the 350 cu in V8 powered RS model while the Z-28 was recorded as having taken only 6.3 seconds to reach 60 mph in a British magazine test. And that was in a used car!

It is a great tribute to the second generation Camaro's overall design that it changed little for the next eleven years. In an ever changing world it is a miracle that the Camaro stayed the same for so long and we have Bill Mitchell to thank for its durable design.

The Ultimate Camaro. Part II

Ponycar sales began to dip dramatically in 1970. 1969 had seen 243,095 Camaros produced and only 124,898 sold, or half the previous year's total. The next three years were even worse; it was as if Camaros, Mustang and their kind had been infected with an incurable disease. Car makers, notably Chrysler, unceremoniously dumped their Ponycars. Ford's Mustang II, introduced in 1974, was a mere shadow of its former self, and GM seriously considered giving the Camaro the bullet after 1972. This also meant Camaro's near twin, Pontiac's Firebird would be discontinued. Fortunately, Chevrolet and Pontiac executives had more faith in their products than GM management obviously had and weren't about to send their cars to the scrap heap. Much determined lobbying ensued to convince the top brass it would be foolish to rid themselves of two of the best cars the company had. Management thought and concurred. The Camaro was saved, the Firebird lived to tell the tale.

What of the Z-28 during this fateful period? The 302 V8 was gone, to be replaced with the 350 cu in V8. As its parentage was Corvette's LT-1 engine it had plenty of go. Fitted with a Holley four barrel carburettor, an aluminum high rise manifold and the old 302s mechanical lifters, the 350 propelled the Z-28 off the line as fast as the 302.

Better handling

The new Z-28 was a superior handling machine to its predecessors, so much so that some car journals claimed it was the best handling American car they'd tested, while others observed that Chevrolet had at last created a true American Grand Tourer. And for the next couple of years that's just what it was, for Chevrolet gave up competition with the loss of Roger Penske to AMC, where he raced in the Javelin.

Suddenly the Z-28 was no more. At the end of 1974, GM decided sales (which were increasing again) weren't enough to justify the car's continued existence. There was also the expense of bringing it into line with the ever increasing emissions controls, and who knew what the future held after the near disastrous 1973 oil embargo? So out went the Z-28, leaving an irreplaceable void in the Camaro lineup.

Starting IROC

At about this time, Roger Penske and others held a roundtable conference. On the agenda was the ingenious, yet quite obvious idea of staging a series of races pitting the world's best racing drivers against each other in exactly the same cars. The drivers would come from all varieties of racing: Formula One, Indy 500, NASCAR and so on. The series of four races would be held on tracks corresponding to the type of racing the participants were normally used to – or not, as the case may be. The race would be known as The International Race of Champions or IROC for short. The first race took place in 1973, the drivers racing identical Porsches. The cars proved less than reliable, so Chevrolet was approached. To cut a long story short, Chevrolet supplied each driver with identical Camaros which were race-prepared by Grand National Stock Car builder Banjo Mathews at his shop in Asheville, N.C.

The Camaros did so well in IROC that they have been used to this day. The resultant publicity spun off nicely in Chevrolet showrooms, so it was considered desirable to revive the Z-28. In mid 1977, the Z-28 once again stood proudly at the helm, an even better car than before.

The new Z-28, although a beauty, wasn't in the same performance league as the earlier cars. The 350 V8 put out a mere 185 bhp (net) and all the racing parts had gone. What had improved however, was the car's handling. Larger wheels and radial tires, stiffer spring rates, a thicker front anti-roll bar and a more flexible one at the rear, gave the car a ride not dissimilar to Europe's best.

Top left: Apart from minor details, 1970-73 Camaros were identical in appearance. A new grille distinguished 1974 models and remained the same until 1978 when urethane front, body colored bumpers and extra length determined it was a new model (General Motors). Bottom left and above: A 1977 LT Coupé with chrome bumpers now in their final year.

End of the road

Little was changed from 1977 to 1981 apart from evolutionary refinements. In fact it is almost impossible to tell one from another. Its ride, handling and surprisingly quick acceleration (0-60 in around 7.5-8.0 seconds), made it a world class car. Apart from giving it independent rear suspension, there seemed little else that could be done to make it better still. After eleven years, a replacement was long overdue, bearing in mind a normal model's life span.

New generation

The third generation Camaro was launched early in 1982. It was sleek; it was aerodynamic; it was beautiful. It took almost seven years to develop but the time spent was more than justified. Battle hardened veterans of the motoring world excused themselves a sharp intake of breath when they saw it for the first time. With its F-body companion, Pontiac's Firebird, color and confidence were brought back after years of gray-suited dullness from Detroit.

Out of three models the top dog was the Z-28. It had a 305 (5 liter) 4 bbl. V8 or the optional 'Cross Fire' throttle body, fuel injected 305 under the hood. This was allied to a standard 4 speed manual transmission. Turbo-Hydra-matic was optional.

A lot of work was done on the Z-28 heavy duty suspension. The front control arms were replaced with MacPherson struts for more rigidity at the rear, and the old leaf springs gave way to coils, but the rigid axle was retained, though its location was fixed by two trailing links and the incorporation of a full width Panhand rod. Axle wind up was cancelled by a long torque arm anchored at the transmission. Front and rear, the Z-28 employed thicker anti-sway bars than the base Camaros.

To beat Mustang

Weighing a hefty 3031 lb, more if you added all the options, the Z-28 had had its length cut by 10 inches to 187.8 inches, and its wheelbase from 108 to 101 inches. Its trim new size

Top left: 1977 Camaro bumper/grille assembly contrasts with 1978 version, top right. Left: 1978 Berlinetta script looks the part, even if it is plastic. By now more and more plastic was being used in an effort to cut weight and save money. At least Chevrolet used new steel after two or three years building Camaros from cheaper, recycled metal. That's why it's difficult to find early to mid seventies Camaros that don't look like brown lace curtains!

and aerodynamic styling (drag co-efficient was 0.37), however, was totally lost on the anemic 165 bhp V8, engine. Consequently the rapid 1982 Mustang GT humiliated it whenever the two chanced to meet.

Chevrolet doesn't take kindly to insults from Ford, and the following year the engineers dropped a 'High Output' 5 liter engine into the Z-28. This engine, rated at 190 bhp, was developed from a hot racing engine Chevy had built. This did the trick; the Zee was able to hold its own, and invariably beat the Mustang GT. One journal recorded a 0-60 time of 6.7 seconds. This was the sixties brought back to life; a million memories of yesterday were turned into realities. Performance was back with a vengeance!

The Zee didn't rest on its laurels. Back it came in 1984 with subtle suspension improvements. It beat all other contenders for the handling crown by a mile, including Mustang's highly rated new SVO.

Into 1985 and the Z-28s power output is up to 215. Torque is a healthy 275 lb-ft at 3200 rpm. A five speed Borg Warner manual is standard, with 4 speed automatic as a desirable option. Even more desirable is the optional 4 wheel power disc brake system coupled with limited slip differential.

Superlative car

If the 1985 Zee is a handling gem, then the new IROC-Z is, as Chevrolet proudly admits, the ultimate. Essentially a $600-plus performance and handling package, the IROC-Z was inspired by the IROC racing Camaros, even though there are few similarities under the skin. (The specially built and prepared IROC racing Camaros have a tubular frame to start with.) This is not to say the IROC-Z is lacking for anything – its prototype beat all comers in a 24 hour race at Nelson Ledges.

One runs out of superlatives when discussing the IROC-Z – or the new Z-28, for that matter. These are very definitely World Class Sports Cars, equal to the very best, regardless of price. It's going to be a hard act to follow in the future.

Main picture: A stock 1981 Camaro Z-28. The 5.7 liter (350 cid) V8 was – and is – the largest engine available. Standard power plant on '81 Camaros was the 5.0 liter (305 cid) V8. Insets: 1979 Camaro in racing guise. Camaros have a good track record in competitive events and can be found in the pits of the world's racing venues.

Z28

These pages: Several views of the attractive 1981 Z-28 beautifully finished in charcoal metallic. This was the last year of the 2nd generation series and had gone as far as it could in terms of refinement and handling.

Z28

A thorough sportscar, lithe, handsome and refined. The Z-28, a car that has everything for the man that hasn't. In other words it's affordable and in 1981 represented tremendous value for money yet looked like a million dollars.

Radial T/A
BFGoodrich

AIR INDUCTION

These pages: A fully optioned '81 Z-28 a-rarin' to go. Its 350 cid 4 Bbl. 185hp V8 could take the car to 60 in 7.5 secs., quick by any standards, especially in 1981. Removable glass 'T-tops,' electric windows, air conditioning, cloth upholstery, Am/Fm stereo radio cassette, automatic transmission, rear window 'sports slats' (louvers) and functional solenoid operated hood air induction scoop make this Camaro a most desirable creature. And, as the brochure says, the only thing missing is you!

A 1980 Z-28, though car spotters would be hardput to spot the difference between 1980-81 models. Its not that easy to spot the difference between 78, 79 models either but simulated fender louvers and hood scoops give the game away. 1980/81 scoops were for real!

Radial T/A
BFGoodrich

TO CONTINUE AND CONSOLIDATE

The second generation Camaro proved it was unnecessary to rely on annual face-lifts or a new body every three years. Apart from a revised front end and wrap around tail lights in 1974 and a new rear window in 1975, the body remained unaltered for eleven years. As time went on, the car's popularity – and sales – increased. As children grew into adulthood the Camaro was always there; it achieved an identity and people were able to identify with it.

Those eleven years were a period of constant evolution. Subtle detail and mechanical modifications raised the Camaro to new heights of road ability and sophistication. Most of the engineering emphasis in the seventies was directed towards handling rather than buccaneer performance which was severely restricted by goverment legislation. The Camaros became Grand Tourers in every sense of the word. The gut wrenching, blood and thunder 'wheelie' days were a thing of the past. Even so, there was

nothing to stop an enterprising mechanic or teenage enthusiast from dropping in a bigger engine and taking the result down to the drag strip. Look at the fatter than usual tires, listen to that exhaust and you know that ain't no standard Camaro!

Four models

Camaros came in four models, during the seventies, all based on the same coupé body. These were the base sport coupé, Rally Sport, LT and Z-28. In 1979 the LT became 'Berlinetta' and was the luxury Camaro. Engines comprised one six and three V8s. The 229 cid., V6 was standard in all but the Z-28, which came with the 350 V8.

The third generation Camaros debuted in 1982. As before, the F-body was shared with Pontiac's Firebird, although there were enough differences to distinguish the two.

As with every other car produced by GM, the new Camaros were downsized some 10 inches, with a wheelbase loss of 7 inches. Even so, the car's curb weight was still on the high side, (2875 lb for the base sport coupé; over 3000 lb for the Z-28).

Development of the 1982 Chevrolet/Pontiac F-cars took nearly seven years, a long time in any car maker's book. The end product more than justified itself for the overall effect was stunning and its driveability was even more so. *Motor Trend* liked the Camaro so much that it was selected for its prestigious 'Car of the Year' award. It also was chosen as the Indianapolis 500 Pace Car for 1982. Not that anyone disagreed: on the contrary. The new Corvette was still a year away and the Camaro was hailed as the best looking car to come out of Detroit in years.

Above left and right: A drizzly day in northeast Indiana and a 1979 Sport Coupé powered by a 4.1 liter (250 cid) L6 – the last year for Chevy's venerable in-line unit: it was replaced by a 229 cid V6 in 1980.

Low performance

Not only stunning to look at, the Camaro was up there with the best of them in terms of ride and handling. Test after test proved it the best handling car built in America and close to some of Europe's over-rated, ultra-expensive toys. The one area that let the Camaro down however, was performance. Even with its top engine option it was a non-starter compared to the furious potential of some of its European rivals.

Model line up consisted of the base Sport Coupé, the luxurious Berlinetta and the Z-28. With an eye on the EPA fuel economy averages, the base Sport Coupé's standard engine was a 2.5 liter 151 cid., four with electronic fuel injection. This was mated to a four speed manual transmission, but hardly the way to go if performance was the choice. A little more 'go' could be had from the Berlinetta's standard 2.8 liter 173 cid., V6 with 5 speed manual overdrive. Top engine was the 5.0 liter 305 cid., in either 4 bbl or electronic fuel injection mode.

Much work had been done on the suspension and chassis/body structure. As already noted, the old A-arms were replaced by MacPherson struts in the front, and to save weight, all drive train and suspension components were bolted onto the main structure, dispensing with the old front sub-frame entirely.

We need not repeat the rest of the details already described in the previous chapter except to say that the Sport Coupé or Berlinetta could be ordered with the optional, sportier F-41 suspension: the Z-28 had its own system.

New systems

1983 saw the arrival of the high output 5.0 liter 305 cid., V8 for the Z-28. Power was up to 190 cid. By 1985 this engine was slicker still and boasted 215 hp at 4400 rpm. With its new Tuned-Port Fuel Injection System and 4 speed automatic transmission (introduced in 1984), its 0-60 time of 7.0 seconds recalls the rubber burning days of the sixties. The 5 speed manual is also available, which might cut acceleration times

even further. With performance back, plus handling to match, the European sports car makers can no longer afford complacency.

The Berlinetta still features the Multi-Port Fuel Injected V6 and the Sport Coupé stays with 4 cylinders. It can be ordered with V6 or the 4 bbl. V8 if the buyer so wishes.

World class standard

Sophisticated, technically advanced and world class – that's Camaro today. It's unlikely Chevrolet can improve much on the current models, unless it adds independent rear suspension, which would appear doubtful. What is in store for the future then? It's hard to say for sure, without speculating a little. Bearing in mind that Corvette, Fiero, Eldorado, Seville, Tornado and Riviera all boast IRS, the next Camaro generation is sure to follow suit. The car may shrink in size, have front wheel drive and possibly follow Pontiac's exciting Fiero in being mid-engined.

Whatever Camaro does you can be sure of one thing: It will remain the 'ultimate' regardless of price.

These pages: The 3rd generation Camaro. It's not often a model can eclipse what went before but Chevrolet managed it with the all new '82 Camaro, 'A whole new car – A lot of new technology', boasted Chevrolet.

GOODYEAR
Z28

The 1982 Z-28. Wind tunnels and computers determined the car's final shape. Drag coefficient was 9% better than the '81 model, at .374.

WANDER
2G 2331
Cross-Fire Injection
Z28

These pages: Called Dark Gold Metallic, the color blends tastefully with the swoopy, sensual shape of the 3rd generation Z-28. Note orange rear indicator lights, a safe and good idea from Europe.

These pages: Whichever way you view it, the 3rd generation looks good, an acknowledgment of excellent design. Here's an '84 Z-28 complete with T-tops and sports slats. At around $15,000 fully loaded, there's little to touch it.

EAGLE GT
EAGLE GT

GOODYEAR
EAGLE GT

This is an '83 Z-28 equipped with cross-fire injection. Note stylish wheels and expensive looking interior. The days of annual change and planned obsolescence have gone; the money isn't around for that sort of wasteful thinking anymore so it's difficult to spot the year a model was made.

These pages: The Ultimate Camaro, the 1985 IROC-Z. A $650 option package, it transforms the already world beater. 0-60 in 7.2 seconds with the optional High Output Tuned Port Fuel Injected 5.0 liter V8 belies the IROC's racing heritage. Insets: Oddly, the hood louvers don't function and it's a shame the recessed lights don't have glass covers, the fault of state-by-state antiquated laws. A rear spoiler completes the picture.

CAMARO

CAMARO
CAMARO

Opposite page and below: The IROC-Z looks good standing still, even better in full flight. Bottom: How the racing IROC Camaro compares with production Z-28: wheelbase is 105 inches, length 191.4 for the IROC compared with 101.00 inches, 187.8 inches for the Zee. The IROC racing car sits 1.3 inches lower than the standard production car but other specifications are remarkably similar (General Motors).